Sheet Metal Mathematics & Geometry Development

Reference Text Book
Edition 1

By

Paul Lawrence

Sheet Metal Mathematics & Geometry Development - Reference Text Book Edition 1
Copyright © Paul Lawrence, 2015
ISBN-13: 978-1530444830

Cover design: Helen Downs www.creativedevotions.net
ISBN-10: 1530444837

List of Figures

Paul Lawrence was born in New South Wales, Australia. Paul's writing draws on his experiences serving in the Australian Army, as well as his knowledge as a qualified Sheet Metal Tradesman and his studies in Applied Science (Cert. App. Science – CAD) and Engineering (Dip. Eng.). Paul's experience dealing with his personal illness, schizophrenia, for which he spent some time in a mental hospital, also provides creative fuel for his creativity.

He is creative in the fields of writing, art, pure math and geometry. He is an inventor and artist.

"Listen to advice and accept instruction, and in the end you will be wise."

Proverbs 19:20

1. CONVERTING METRIC DIMENSIONS TO IMPERIAL FRACTIONS

Legend:

25.4mm = 1 Inch

$\frac{64}{64}$ = 1 Inch

12 Inch = 1 Foot

3 Feet = 1 Yard

5280 Feet = 1 Mile

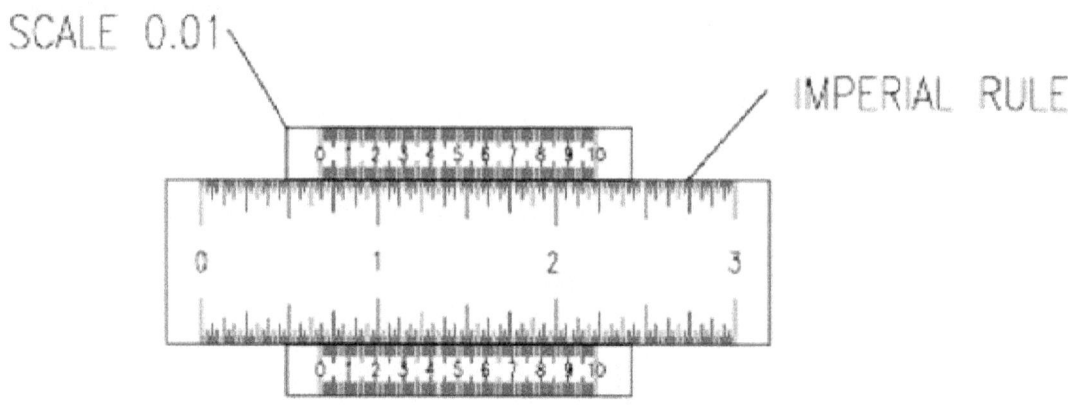

Figure 1 64th Imperial Rule with 0.01 (hundredth) scale

1 Meter = 1000 Millimeter

1000 Meters = 1 Kilometer

This diagram of a 64th with a hundredth of a 64th was created with CAD. It illustrates that you can measure to a hundredth of a 64th and therefore increase the accuracy of conversion from metric to measurable imperial measurement. This reduces errors when converting from the imperial measurement back to metric. This scale illustrates this normal maths equation to give a functional real value to the final answer, which can be measured to one hundredth of a 64th as shown in **1.0 Maths Equation to Hundredth of 64th.**

This maths equation with Pythagoras illustrates that normal maths can be used and that it gives an answer in imperial inches and decimals, which can now be converted to a measurable value that allows for improved accuracy.

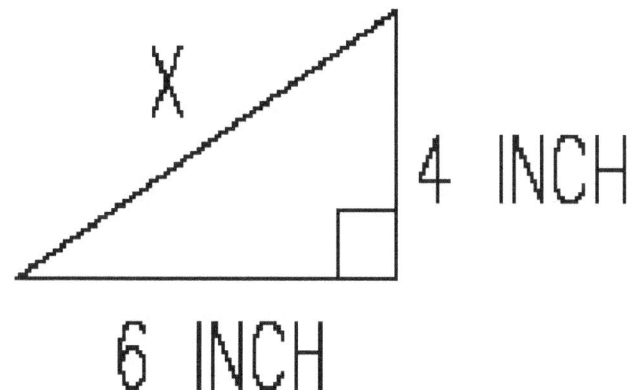

$X = \sqrt{6^2 + 4^2}$

$\quad = \sqrt{36 + 16}$

$\quad = \sqrt{52}$

$\quad = 7.211102551"$

Therefore:

$$7" + \left(\frac{(\,0.211102551 \; x \; 64\,)}{64} \right)$$

$$= 7" + \frac{13.51056326}{64}$$

$$= 7" + \frac{13.51}{64}$$

$$= 7" \frac{13.51"}{64}$$

1.1. Converting millimeters to imperial fractions

Example 1: Convert 777mm (0.777M) to Feet Imperial Fraction

- Divide 777mm by 25.4 to give you an imperial decimal value of 30.59055118.
- Divide 30 by 12 to give feet: 2 feet and 6 inch remaining.
- Multiply 0.590551181 by 64.
- The 64th then becomes a denominator with the value of 37.795 as the numerator: 37.795/64th.
- Round off 37.795 to the nearest hundredth whole number: 37.8/64.
- Therefore, the imperial fraction reads: 2' 6" 37.8/64".

To check the imperial fraction:
- Multiply 2' by 12 and add 6": 30"
- Divide 37.8 by 74: 0.590625.
- Add 30 to 0.590625: 30.590625.
- Multiply 30.590625 by 25.4 gives you 777.001875 mm – which is accurate allowing for error in rounding off.

0.777M x 1000 = 777mm

$$= \frac{777mm}{25.4} = 30.59055118"$$

$$= 30" + (\frac{(0.590551181 \times 64)}{64})$$

$$= \frac{30"}{12} + \frac{37.79527559}{64} \qquad \frac{30}{12} = 2.5 \text{ Feet}$$

$$= 6" + \frac{37.8}{64} \qquad <=> \quad 30" - (12 \times 2) = 6"$$

$$= 2' + 6" + \frac{37.8}{64} \qquad 2 \text{ Feet}$$

$$= 2' 6" \frac{37.8"}{64}$$

Example 2 Convert 7mm to Imperial Fraction

$$\frac{7}{25.4} = 0.275590551$$

$$= (\frac{(0.275590551 \times 64)}{64})$$

$$= \frac{17.63779528}{64}$$

$$= \frac{17.64"}{64}$$

Example 3: Convert 77mm to Inch Imperial Fraction

$$\frac{77}{25.4} = 3.031496063"$$

$$= 3" + \left(\frac{(0031496062 \times 64)}{64}\right)$$

$$= 3" + \frac{2.015748031}{64}$$

$$= 3" + \frac{2.02}{64}$$

$$= 3" \; \frac{2.02"}{64}$$

Example 4: Convert 7.65M to Yard Imperial Fraction

$$7.65M \times 1000 = 7650mm$$

$$= \frac{7650}{25.4} = 301.1811024$$

$$= 301" + \left(\frac{(0.181102362 \times 64)}{64}\right)$$

$$= \frac{301}{12} + \frac{11.59055118}{64} \qquad\qquad \frac{301}{12} = 25.08333333 \text{ Feet}$$

$$= 1" + \frac{11.59}{64} \qquad\qquad 301" - (12 \times 25) = 1"$$

$$= 8 \text{ Yards} + 1' + 1" + \frac{11.59}{64} \qquad <=> \quad \frac{25 \text{ Feet}}{3} = 8.333333333 \text{ Yards}$$

$$= 8 \text{ Yards } 1' \; 1" \; \frac{11.59"}{64} \qquad\qquad 25 \text{ Feet} - (3 \times 8) = 1 \text{ Foot}$$

$$8 \text{ Yards}$$

Example 5: Convert 77.7KM to Mile Imperial Fraction

$$77.7KM \times 1000 = 77700M$$

$$= 77700M \times 1000 = 77700000mm$$

$$= \frac{77700000}{25.4} = 3059055.118$$

$$= 3059055" + \left(\frac{(0.1181102 \times 64)}{64}\right)$$

$$= \frac{3059055}{12} + \frac{7.5590528}{64} \qquad\qquad \frac{3059055}{12} = 254921.25 \text{ Feet}$$

$$= 3" + \frac{7.56}{64} \qquad\qquad 3059055" - (12 \times 254921) = 3"$$

$$= 48 \text{ Mile} + 493 \text{ Yards} + 2' + 3" + \frac{7.56}{64} \quad <=> \quad \frac{254921 \text{ Feet}}{5280} = 48.28049242 \text{ Mile}$$

= 48 Mile 493 Yards 2' 3" $\dfrac{7.56"}{64}$

254921 Feet - (5280 x 48) = 1481 Feet

$\dfrac{1481\ Feet}{3}$ = 493.6666667 Yards

1481 Feet - (3 x 493) = 2 Feet

493 Yards

48 Mile

2. CALCULATING THE LENGTH OF AN ARC

In manufacturing, it is useful and advantageous to be able to calculate accurately the length of an arc, to achieve quality tolerance. Here is an example of how it is done: (See Figure 2)

Working on the mean diameter, draw an arc (90mm radius) at an angle of 58 degrees. To calculate the length of material to produce this item, use the following equation:

$$\text{Length of arc} = \frac{Mean\ Diameter\ \text{x}\ \pi}{360°}$$

$$= \frac{\text{Ø}180\ \text{x}\ \pi}{360°}$$

$$= 1.5708\text{mm}\ /\ °$$

Therefore

$$= \frac{1.5708mm}{1°}\ \text{x}\ 58°$$

$$= 91.1064\text{mm}$$

Calculation:

- Mean diameter (180mm) multiply pi and divide by 360 degrees will give you 1.571mm/degree.
- Multiply 1.571mm/degree by 58 degrees, cancel out the degrees and the final answer will be a length of 91.106mm.

The benefit of this calculation is that you achieve good accuracy to meet tolerance expectations.

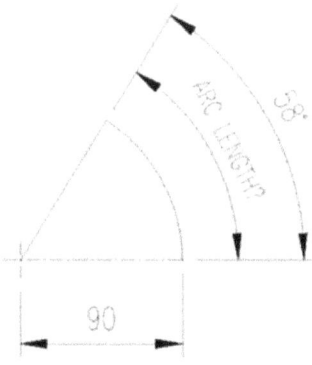

Figure 2 Calculating the length of an arc

3. CALCULATING THE LENGTH OF A SPIRAL AROUND A CYLINDER

The length of a spiral is calculated in order to ascertain the amount of material required to wrap around a cylinder.

3.1. One pitch spiral

Working on the mean dimension, draw a front elevation of a cylinder with a diameter of 90mm and a length of 100mm. Divide the front elevation into twelfths. Division of the length is achieved by using an equal spaced dimension. Where the lines of intersection occur, draw a curve passing through these points. Project the equally spaced lines onto plan elevation and draw a curve at the intersection of these points. (See Figure 3 for diagrams.)

Developing a flat pattern will provide an illustration of the equation to obtain the length of a spiral. Flat pattern length is calculated by multiplying the mean diameter by pi (90mm x pi = 282.744mm. Round off to nearest half millimeter: 283mm). Divide this into twelve equal parts. Width is 100mm; divide this also into 12 equal parts. Draw a line at the points of intersection. The following formula is used to obtain the length:

Spiral Length $= [\sqrt{(Mean\ Diameter \times \pi)^2 + (Pitch^2)}] \times$ Number of Pitches

Calculation

- In brackets, mean diameter (90mm) multiplied by pi and squared will give you 79943.796mm rounded off.
- The pitch (100mm) is squared to give you 10000mm.
- Add the 79943.796mm to 10000mm to give you 89943.796mm rounded off.
- Find the square root of 89943.796mm which will give you 299.906mm rounded off.
- Multiply 299.906mm by the number of pitches (1) to give you a figure of 299.906mm for the length of a spiral.

The benefit of this calculation will be to produce a length of material to complete this job and to know how much material is required to purchase.

Spiral Length $= [\sqrt{(Mean\ Diameter \times \pi)^2 + (Pitch^2)}] \times$ Number of Pitches

$= [\sqrt{(\text{Ø}90 \times \pi)^2 + (100^2)}] \times 1$

$= [\sqrt{79943.796 + 10000}] \times 1$

$= [\sqrt{89943.796}] \times 1$

$= 299.906 \times 1$

$= 299.906mm$

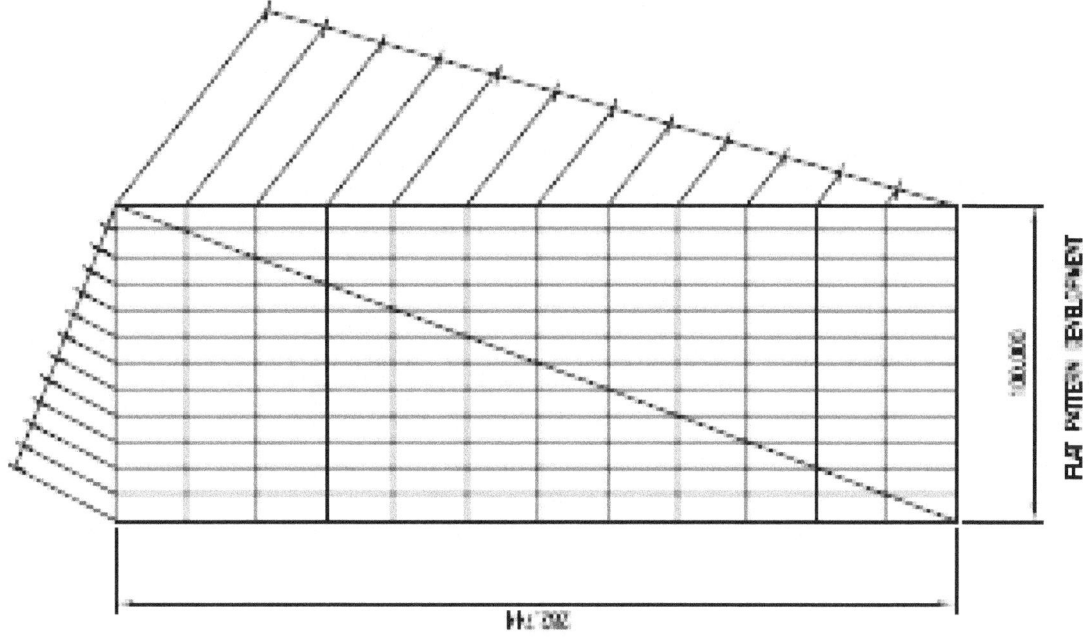

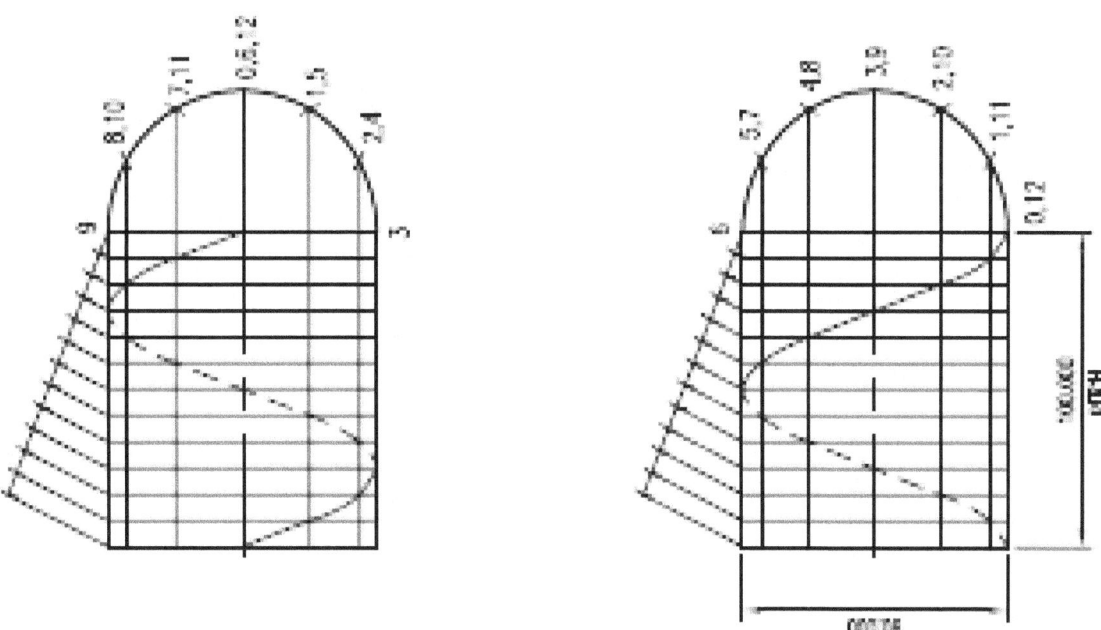

Figure 3 Calculating the length of a spiral around a cylinder

4. DIVIDED LINE METHOD

In the development of truncated cylinders, the divided line method can achieve greater accuracy than using calculator values for the twelfth in flat pattern development. This method eliminates the need for sliding rulers and set squares, which can tend to slip. (See Figure 4 for diagram.)

- Draw a line segment at any length and name the ends 0 and 6.

- Draw any length of line at an angle

- With divider step off a length at equal distance six times on the angle line, marking the points from 0 to 6.

- Draw a line from 6 on the angle line through the point 6 on the line segment.

- With a compass draw a circle with the radius from 6 to 5 and draw circles on point 6 and naming points A, B on line segment.

- At A, B construct a perpendicular line giving point C, G and connecting to point 6.

- On the circles at point 6, where the connecting line crosses the circle, label points D, E.

- With radius D, A draw an arc from point E to find point F and draw a horizontal line from F through 6.

- With radius A, G draw an arc to find H from point F.

- With arc H, 5 draw an arc from G to find point J.

- Draw a line to connect 5 and J giving you point K.

- With arc length 6 to K step off on the line segment giving you four points of K and equal points of six.

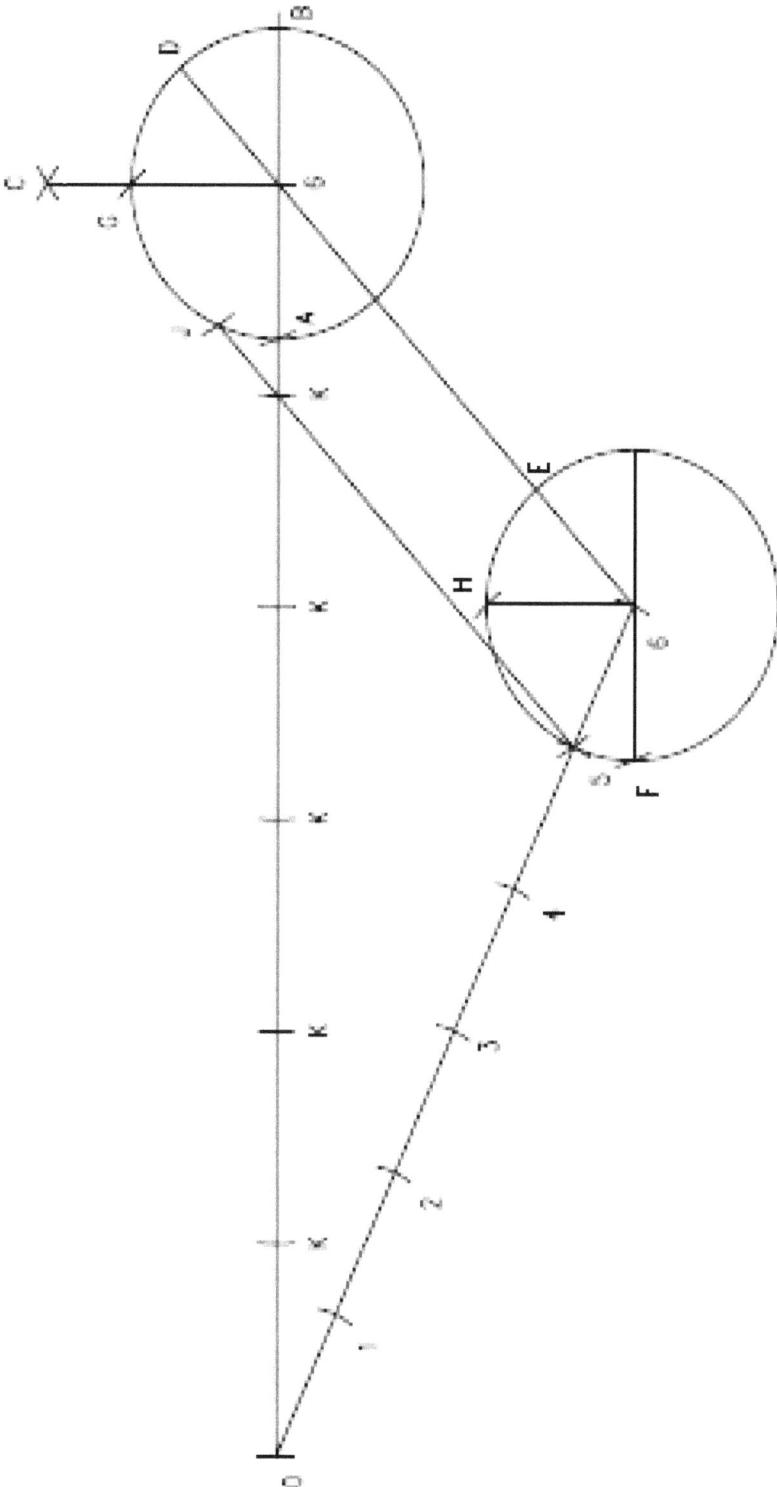

Figure 4 Divided line method

This section is about the development of a cylinder using the **equal spaced divided line method** to obtain greater accuracy.

Application with this method eliminates the error of plan elevation chord limitation, which is that the chord dimension is already in compression and to use this would make your development smaller than it is. (See Figure 5)

Draw a front elevation of a truncated cylinder (diameter 80mm). Measure up 42mm and then draw a truncated 30 degree line. Draw a plane elevation and then divide up into twelfths and project them to the front elevation.

Measure out the length of the flat pattern cylinder: (Mean diameter 80mm x pi = 251.327mm; round off to nearest half millimeter: 251.5mm). Apply the equal spaced divided line method to break up the length into twelve equal spaces. Project from front elevation the truncated cut, which has been divided by the plan elevation projection on to the twelve equal spaces, giving you points of the curve. Draw the curve through the projected points.

The process of a divided line to plan elevation chord length will give an achievable accuracy in developing a product that will meet specified manufacturing tolerances.

Length of Cylinder = Mean Diameter x π

$$= \varnothing 80 \times \pi$$

$$= 251.328\text{mm round off } 251.5\text{mm}$$

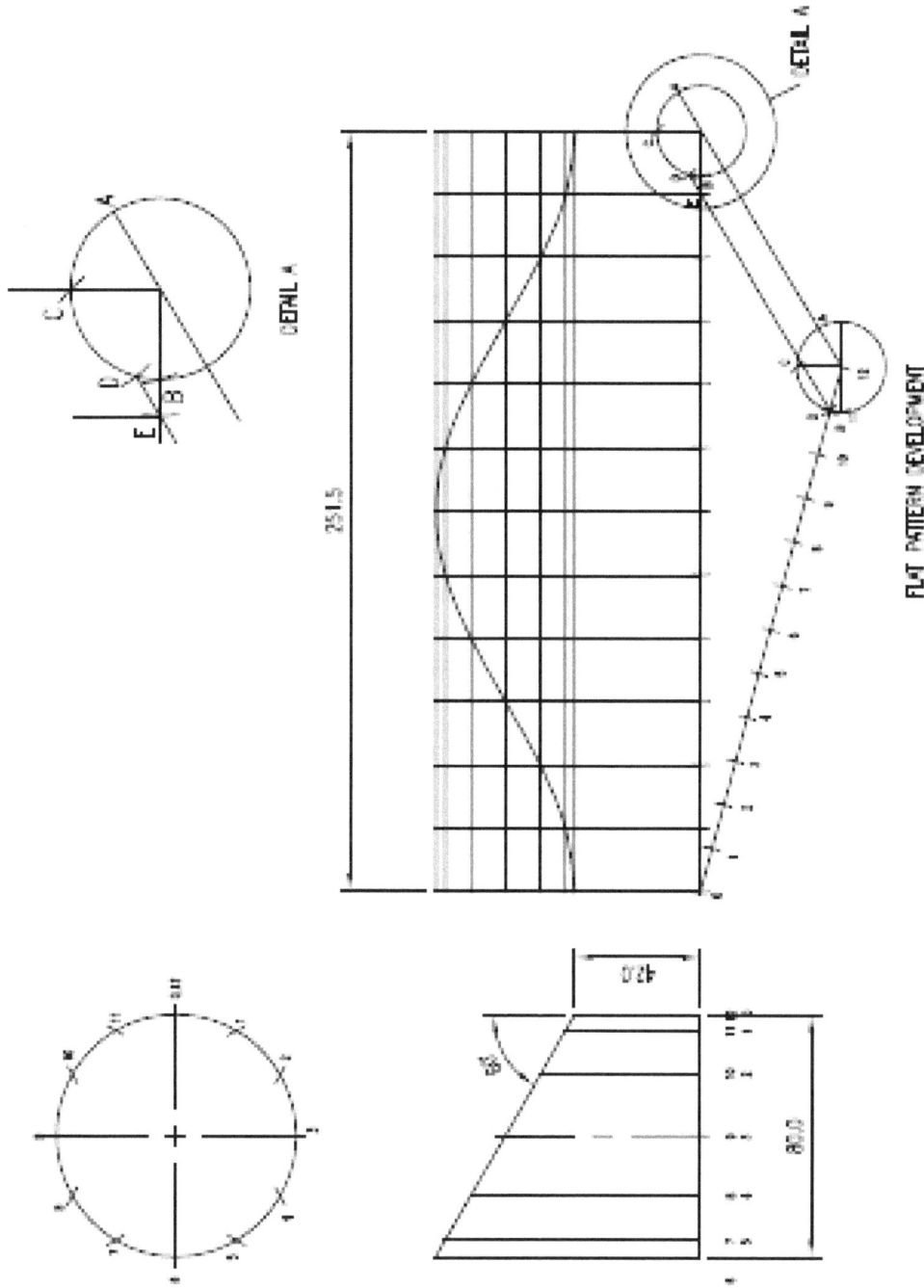

Figure 5 Truncated cylinder development

5. PERIMETER OF ELLIPSE CALCULATION

5.1. Elliptical Concentric Circle Development Method

Using concentric circle developments is an accurate method to construct an ellipse by hand.

Draw a circle with a major axis with a dimension of 100mm and then draw a concentric circle with a minor axis with a dimension of 80mm. Divide both the circles into twelve equal parts numbering 0 to 12. Draw vertical lines from the points 0 to 12 from the outer circle. Draw horizontal lines from the points 0 to 12 from the inner circle. Connecting the point of intersection of vertical and horizontal lines, draw an arc through these points to obtain an ellipse. (See Figure 6)

This is an accurate method to construct an ellipse to a tolerance of plus or minus one millimeter.

Calculation

- In brackets, mean diameter one (100mm) divided by 360 degrees plus mean diameter two (80mm) divided by 360 degrees will give you 0.5mm/degree.

- Multiply the 0.5mm/degree by pi to give you 1.571mm/degree.

- Multiply the 1.571mm/degree by 90 degrees divide by 2 to give you 70.686mm/degrees.

- Multiply the 70.686mm/degrees by 4 to give you an answer 282.744mm/degree. Put all this over a denominator of 1 degree.

- Cancel out the degrees and the final answer is 282.744mm for a perimeter of an ellipse.

Perimeter of an Ellipse $= [[(\frac{D1}{360°} + \frac{D2}{360°}) \times \pi] \times (\frac{90°}{2})] \times 4$

$$= [[(\frac{Ø100}{360°} + \frac{Ø80}{360°}) \times \pi] \times (\frac{90°}{2})] \times 4$$

$$= ((0.5mm/° \times \pi) \times 45°) \times 4$$

$$= (1.571mm/° \times 45°) \times 4$$

$$= 70.685mm \times 4$$

$$= 282.743mm$$

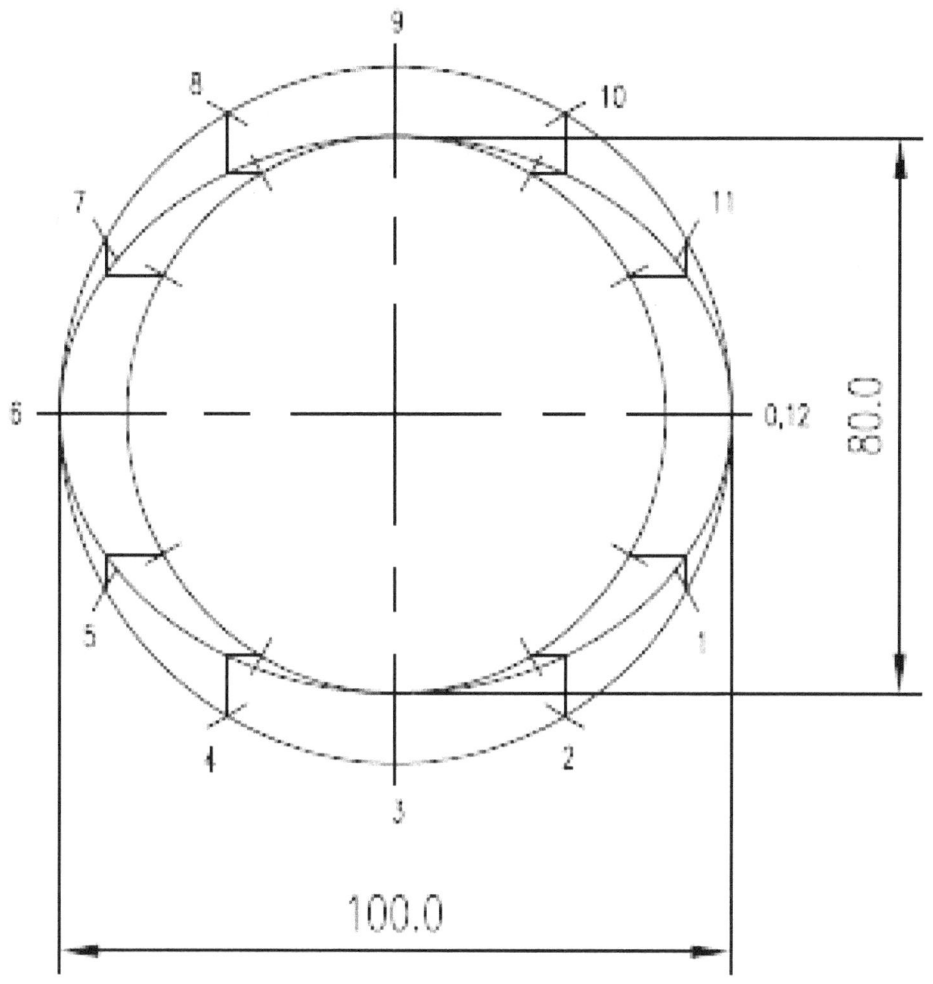

Figure 6 Perimeter of ellipse calculation

6. RIGHT CIRCULAR CONE PATTERN ANGLE

The right circular cone pattern angle is pure mathematics that explains and proves the pattern angle in the development of a cone. With drawing information you can calculate a pattern angle mathematically, which is more accurate than taking a twelfth from a drawn plan elevation because this measurement is under compression, which will give you a smaller cone development. Use mean dimensions.

Calculation of Pattern Angle

- (Slant height x 2 x pi / 360 degree) x unknown angle = (Diameter x pi /12)
- (100 x 2 x 3.1416 / 360) x Unknown angle = (80 x 3.1416 / 12)
- 1.746mm/degree x unknown angle = 20.944mm
- Transpose (cancel out) the degree on both sides of the equal sign.
- Transpose (cancel out) the 1.746mm on both sides of the equal sign.
- Transpose (cancel out) millimeter on both sides of the equal sign to give you degree.
- Divide 20.944 degree by 1.746 to give you 12 degree – the unknown angle.
- Multiply 12 degree by 12 segments will give you the pattern angle of 144 degree.

Pattern Angle = $\left(\dfrac{SHT \times 2 \times \pi}{360°} \right) \times X° = \left(\dfrac{D \times \pi}{12} \right)$

$= \left(\dfrac{100 \times 2 \times \pi}{360°} \right) \times X° = \left(\dfrac{\varnothing 80 \times \pi}{12} \right)$

$= \dfrac{1.745329252mm}{1°} \times X° = 20.944mm$

$= \dfrac{1.745329252mm}{1°} \times X° = 20.944mm \times 1°$

$= \dfrac{1.745329252mm}{1.745329252mm} \times X° = \dfrac{20.944mm/°}{1.745329252mm}$

$= \qquad\qquad X° = 12°$

Therefore

$12° \times 12 = 144°$

Calculation of Chord

- Twelfth segment zero to one will give you a right angle triangle with the hypotenuse of 100 and an angle of six degree. (See Figure 7)
- Using trigonometry sine 6 degree x 100 = 10.453mm

Multiply 10.453mm x 2 = 20.906mm chord length.

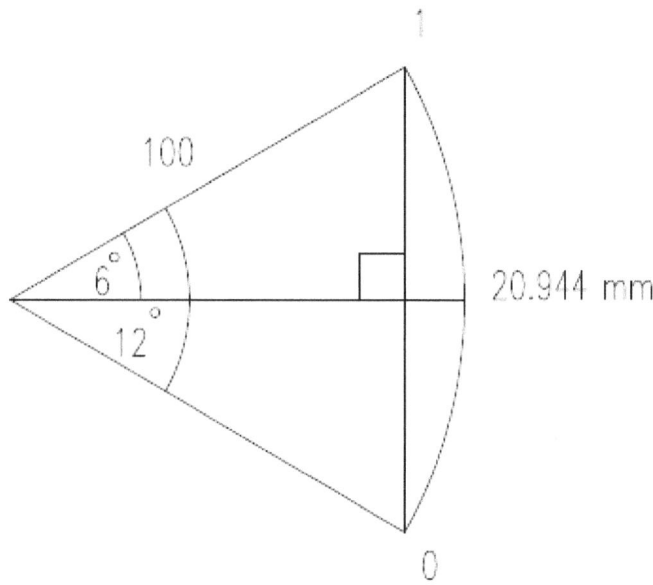

Figure 7 Chord Calculation

Chord = (Sine 6° x 100) x 2 = 20.906mm

Construction

With a divider or trammel, scribe an arc with the slant height as the radii. Pick a point on the arc as 0 and draw a line back to the center point. Place a protractor on the line and mark off 144 degree. Draw a line joining the angle back to the center point. Another option is to measure off the twelfth 20.906mm rounding off to the nearest half a millimeter and scribe the distance with dividers along the drawn arc twelve times. Both methods will give you a development pattern that can be manufactured to specified tolerance. (See Figure 8)

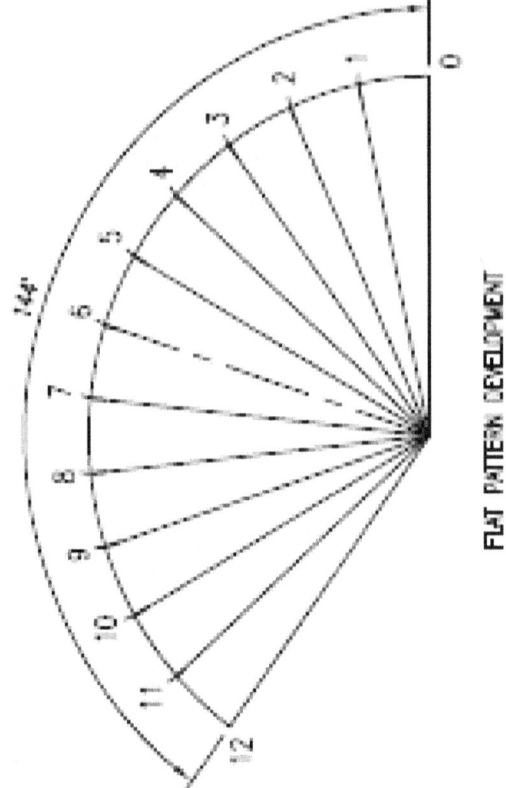

FLAT PATTERN DEVELOPMENT

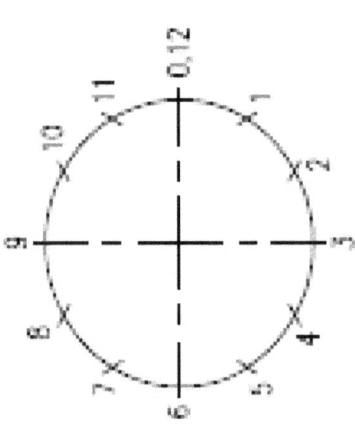

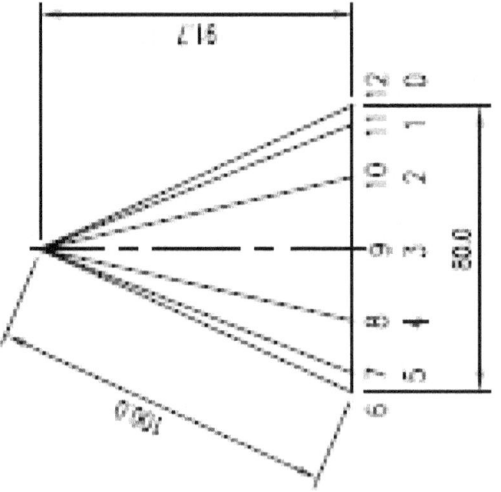

Figure 8 Construction of right circular cone pattern angle

7. SQUARE TO ROUND MATHEMATICAL DEVELOPMENT

Development of a square or rectangle to round can be achieved by geometry or mathematics. To develop with mathematics you have to calculate all true length first and then the chord length. Obtaining the chord from drawn geometry is not a good practice as this method uses the dimension under compression, and to do so will produce a smaller pattern, compressing it once again. The mathematics will bypass the need to draw a 1:1 scale of plan elevation and height to find the true length using geometry. Note that with a material thickness above 5mm you use mean dimension, and under 5mm you use inside dimension. (See Figure 9)

Calculation of True Length of a Square Using Twelfth

Using a standard drawing, or sketching one, show the dimensions – length of the sides, circle diameter and height.

- Sketch in break lines and number the points illustrated in the drawing.
- Find length A0, F12: (200 – 80 / 2) = 60
- Use Pythagoras: sqrt (60 sq + ht 100 sq) = true length 116.619
- Find length B0, B3, C3, C6, D6, D9, E9 E12: sqrt (60 sq + 100 sq) = 116.619
- sqrt (116.619 sq + ht 100 sq) = true length 153.623
- Find length B1, B2, C4, C5, D7, D8, E9, E12: (360 degree / 12) = 30 degree
- Draw in triangle Z, 11, X: (Sine 30 degree x rad 40) = 20
- (cos 30 degree x rad 40) = 34.641
- Triangle 11, Y, E: sqrt (rad 40 + 60 – 34.641 sq) + (100 – 20 sq) = 103.305
- sqrt (103.305 sq + ht 100 sq) = true length 143.778

Step 1 Length A0, F12 = $\frac{200-80}{2}$ = 60

$$= \sqrt{60^2 + Ht\ 100^2} = 116.619 \text{ round off } 116.5$$

Step 2 Length $\frac{B0,C3,D6,E9,}{B3,C6,D9,E12}$ = $\sqrt{60^2 + 100^2}$ = 116.619

$$= \sqrt{116.619^2 + Ht\ 100^2} = 153.623 \text{ round off } 153.5$$

Step 3 Length $\frac{b1,b2,c4,c5}{d7,d8,e10,e11}$ = $\frac{360°}{12}$ = 30°

$$= \text{Triangle Z, 11, X} = \text{Sine } 30° \text{ x R40} = 20$$

$$= \text{Cos } 30° \text{ x R40} = 34.641$$

$$= \text{Triangle 11, Y, E} = \sqrt{(R40 + 60 - 34.641)^2 + (100 - 20)^2} = 103.305$$

$$= \sqrt{103.305^2 + Ht\ 100^2} = 143.778 \text{ round off } 144$$

Calculation of Chord

- (diameter x pi / 360 degree) x (360 degree / 12)
- (80 x 3.1416 / 360) x (360 / 12)
- 0.698mm / degree x 30 degree = 20.944mm

Step 4 Chord $= \left(\dfrac{Inside\ Diameter\ x\ \pi}{360°} \right) \text{ x } \left(\dfrac{360°}{12} \right)$

$\qquad = \left(\dfrac{\emptyset 80\ x\ \pi}{360°} \right) \text{ x } \left(\dfrac{360°}{12} \right)$

$\qquad = \dfrac{0.6981317mm}{1°} \text{ x } 30°$

$\qquad = 20.944 \text{ round off } 21$

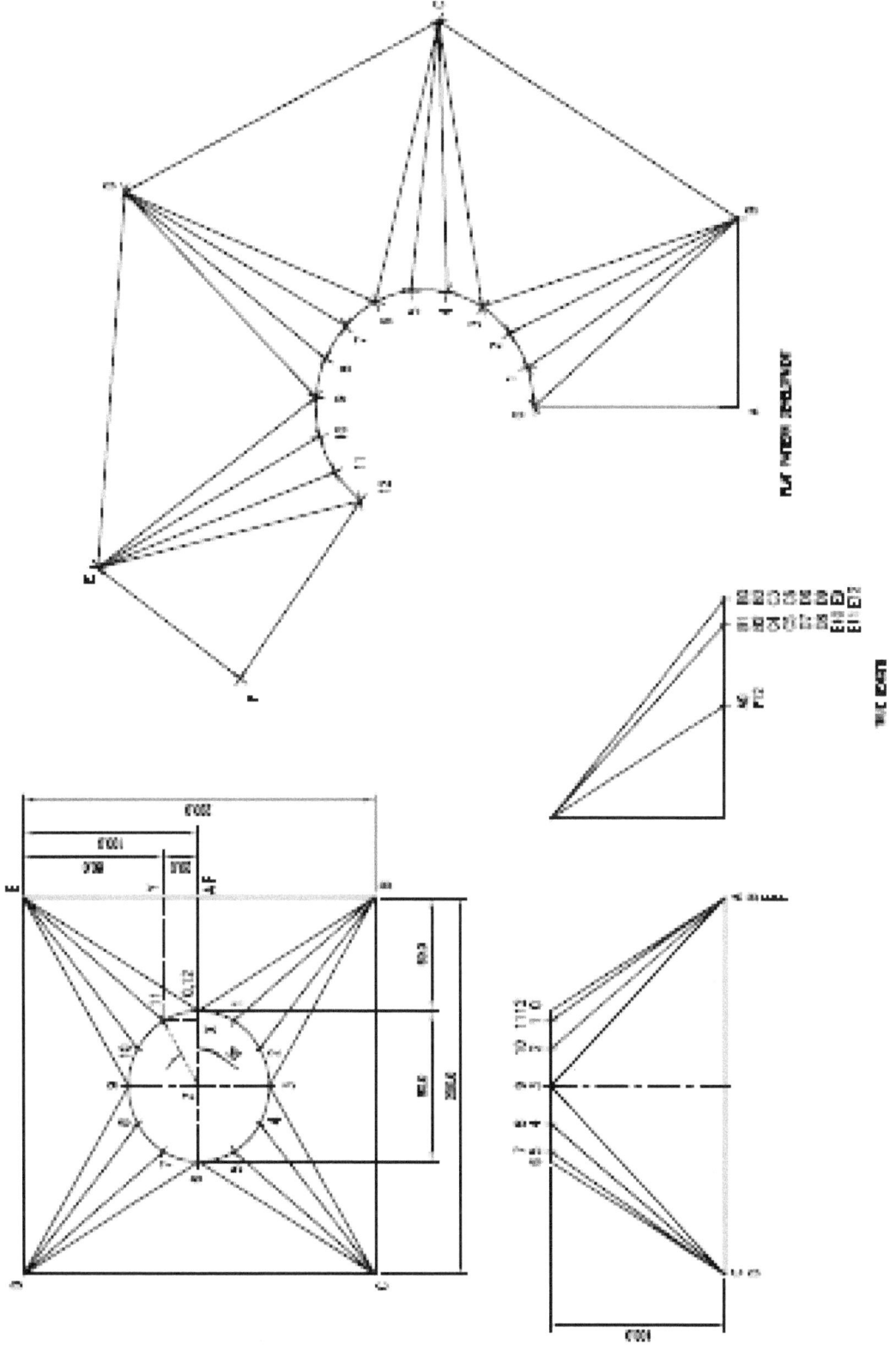

Figure 9 Square to round mathematical development

8. ROUND TO ROUND MATHEMATICAL DEVELOPMENT

Development of a round to round can be achieved by geometry or mathematics. To develop one using mathematics you need to calculate all true length first and the chord lengths. Obtaining the chord from drawn geometry is not a good practice as this method uses the dimension under compression, and to do so will produce a smaller pattern, compressing it once again. The mathematics will bypass the need to draw a 1:1 scale of plan elevation and height to find the true length using geometry. (See Figure 10)

Calculation of True Length of a Round Using Twelfth

- Using a standard drawing, or sketching one, show the dimensions – length of the sides, circle diameter and height.
- Sketch in break lines and number the points illustrated in drawing.
- Find length A0, B1, C2, D3, E4, F5, G6, H7, J8, K9, L10, M11, N12:

$$(200 / 2 - 80 / 2) = 60$$

- Use Pythagoras: sqrt (60 sq + ht 100 sq) = true length 116.619
- Find length A1, B2, C3, D4, E5, F6, G7, H8, J9, K10, L11, M12:

$$(360 \text{ degree} / 12) = 30 \text{ degree}$$

- Draw in triangle Z, X, 1: (Sine 30 degree x rad 40) = 20
- (Cos 30 degree x rad 40) = 34.641
- Triangle X, 1, A: sqrt (rad 40 + 60 - 34.641 sq) +(20 sq) = 68.351
- sqrt (68.351 sq + ht 100 sq) = true length 121.127

Length of Chords

Chord diameter 200

- (diameter x pi / 360 degree) x (360 degree / 12)
- (200 x 3.1416 / 360) x (360 / 12)
- 1.746 degree x 30 degree = 52.36mm

Chord diameter 80

- (diameter x pi / 360 degree) x (360 degree / 12)
- (80 x 3.1416 / 360) x (360 / 12)
- 0.698mm / degree x 30 degree = 20.944mm

Finding True Length Mathematically

Step 1. Length $\dfrac{A0,B1,C2,D3,E4,F5}{G6,H7,J8,K9,L10,N12} = \dfrac{200-80}{2} = 60$

$$= \sqrt{60^2 + Ht\ 100^2} = 116.619 \text{ round off } 116.5$$

Step 2. Length $\dfrac{A1,B2,C3,D4,E5,F6}{G7,H8,J9,K10,L11,M12} = \dfrac{360°}{12} = 30°$

$= \text{Triangle Z, X, 1} = \text{Sine } 30° \times R40 = 20$

$= \text{Cos } 30° \times R40 = 34.641$

$= \text{Triangle X, 1, A} = \sqrt{(R40 + 60 - 34.641)^2 + (20^2)} = 68.351$

$= \sqrt{68.351^2 + Ht\ 100^2} = 121.127 \text{ round off } 121$

Step 3. Chord Diameter 200 $= \left(\dfrac{\textit{Inside Diameter} \times \pi}{360°} \right) \times \left(\dfrac{360°}{12} \right)$

$= \left(\dfrac{\emptyset 200 \times \pi}{360°} \right) \times \left(\dfrac{360°}{12} \right)$

$= \dfrac{0.1.745329252mm}{1°} \times 30°$

$= 52.36 \text{ round off } 52.5$

Step 4. Chord Diameter 80 $= \left(\dfrac{\textit{Inside Diameter} \times \pi}{360°} \right) \times \left(\dfrac{360°}{12} \right)$

$= \left(\dfrac{\emptyset 80 \times \pi}{360°} \right) \times \left(\dfrac{360°}{12} \right)$

$= \dfrac{0.6981317mm}{1°} \times 30°$

$= 20.944 \text{ round off } 21$

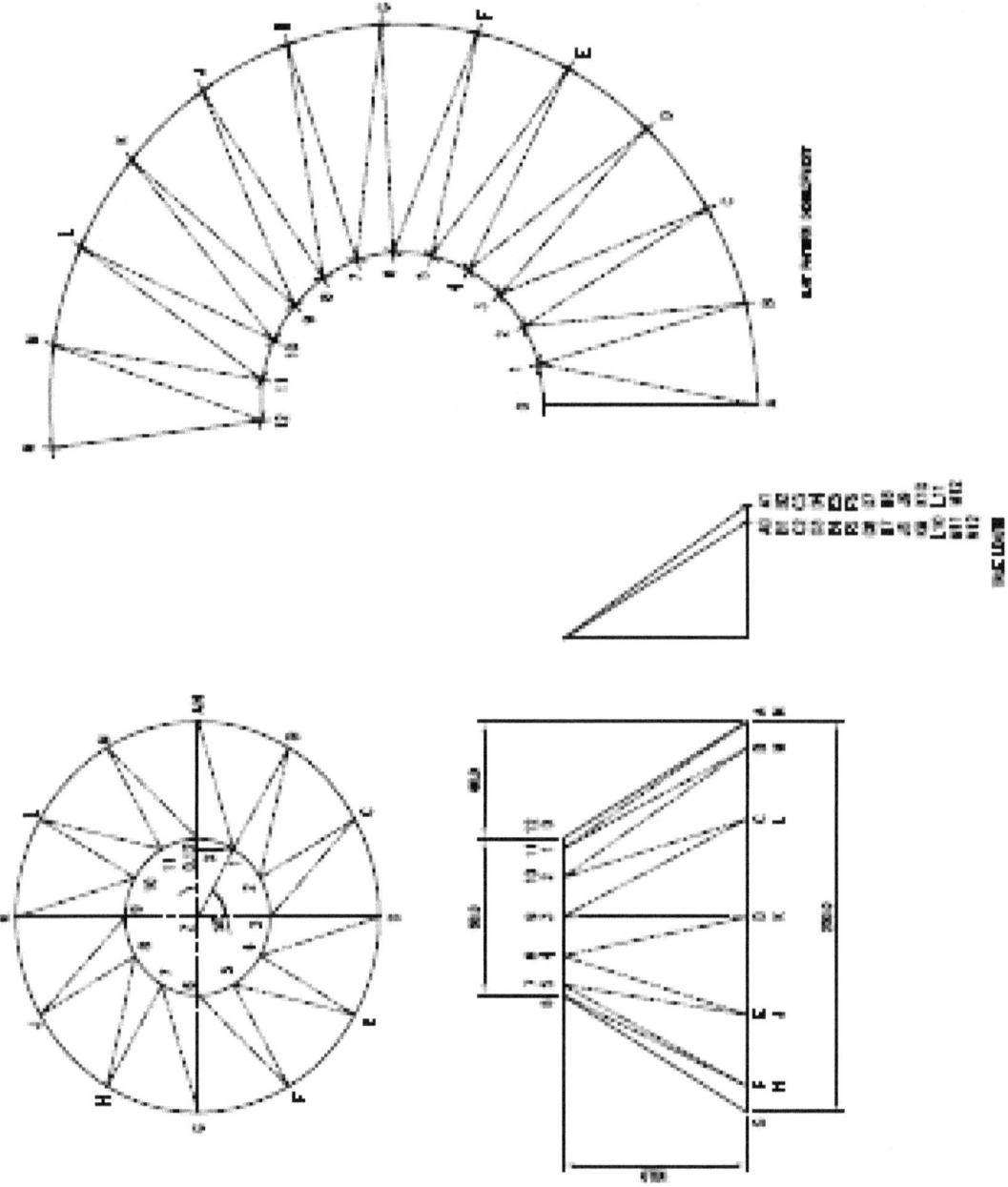

Figure 10 Finding true length of a round using twelfth

9. CUTTING PLANES SPHERE MATHEMATICAL DEVELOPMENT

The flat pattern development of a sphere can be achieved by the cutting-plane method. With this method you will have an approximation, as the value pi is an irrational number. This flat pattern development can also be used to develop a sink bowl corner or urinal bowl corner. Calculation is based on mean (neutral axis), which does not shrink or stretch like the inner and outer surface.

Construction

Working on the mean diameter, draw three circles (diameter 100mm); each circle is for a plane elevation, front elevation and end elevation. Divide the end elevation circle into twenty-four equal points, numbering 0 to 24. Project these points onto the front elevation. At the front elevation number the points starting from the center line moving up the circle 'A' to 'G'. Project these points onto the plan elevation at the center line to create circle cutting plane. Divide the plane elevation into twenty-four equal points and number 0 to 24. At the plane elevation draw circles from the points on the center line to form circles of cutting planes and number 'A' to 'G' in relation to the front elevation. (See Figure 11)

Development of a quarter

- Draw two center lines at right angles to each other. Label the intersection of major/minor axis center line as 3.
- Measure from 3 to 0 in one direction and from 3 to 6 in the other direction: ((Mean diameter x pi) / 4 / 2).
- Measure and mark points 'A' to 'G' on the major center line. This is achieved by using a twenty fourths dimension (((mean diameter x pi) / 360 deg) x 90 deg).
- Use the dividing line method to find points 'A' to 'G' to obtain equal spacing for each dimension of the pervious formula.
- At each cutting plane point 'B' to 'G', use this formula:

((Cos 15, 30, 45, 60, 75 x (100/2) x 2) x (pi/4)) for 'B' to 'F' cutting planes axis.

Draw an arc from 'A (6)' to 'G' and 0 to 'G' connecting all the points on the cutting plane S. Duplicate this procedure to draw the other half on this quarter.

For development of a sink bowl / urinal corner develop one quarter of the sphere pattern. (See Figure 11)

Development of Quarter

Step1. Find quarter 24th of Ø100 = ($\frac{Mean\ Diameter\ x\ \pi}{360°}$) x 90°

$$= (\frac{Ø100\ x\ \pi}{360°}) \times 90°$$

$$= 78.54 \text{ round off } 78.5$$

Step 2. Quarter-diameter half-length at 3.

At A = ($\frac{\frac{Ø100\ x\ \pi}{4}}{2}$) = 39.27 round off 39.5

At B = $\frac{[[[\ Cos\ 15°\ x\ (\frac{Ø100}{2})]\ x\ 2\ ]\ x\ (\frac{\pi}{4})]}{2}$ = 37.94 round off 38

At C = $\frac{[[[\ Cos\ 30°\ x\ (\frac{Ø100}{2})]\ x\ 2\ ]\ x\ (\frac{\pi}{4})]}{2}$ = 34.008 round off 34

At D = $\frac{[[[\ Cos\ 45°\ x\ (\frac{Ø100}{2})]\ x\ 2\ ]\ x\ (\frac{\pi}{4})]}{2}$ = 27.788 round off 28

At E = $\frac{[[[\ Cos\ 60°\ x\ (\frac{Ø100}{2})]\ x\ 2\ ]\ x\ (\frac{\pi}{4})]}{2}$ = 19.635 round off 19.5

At F = $\frac{[[[\ Cos\ 75°\ x\ (\frac{Ø100}{2})]\ x\ 2\ ]\ x\ (\frac{\pi}{4})]}{2}$ = 10.163 round off 10

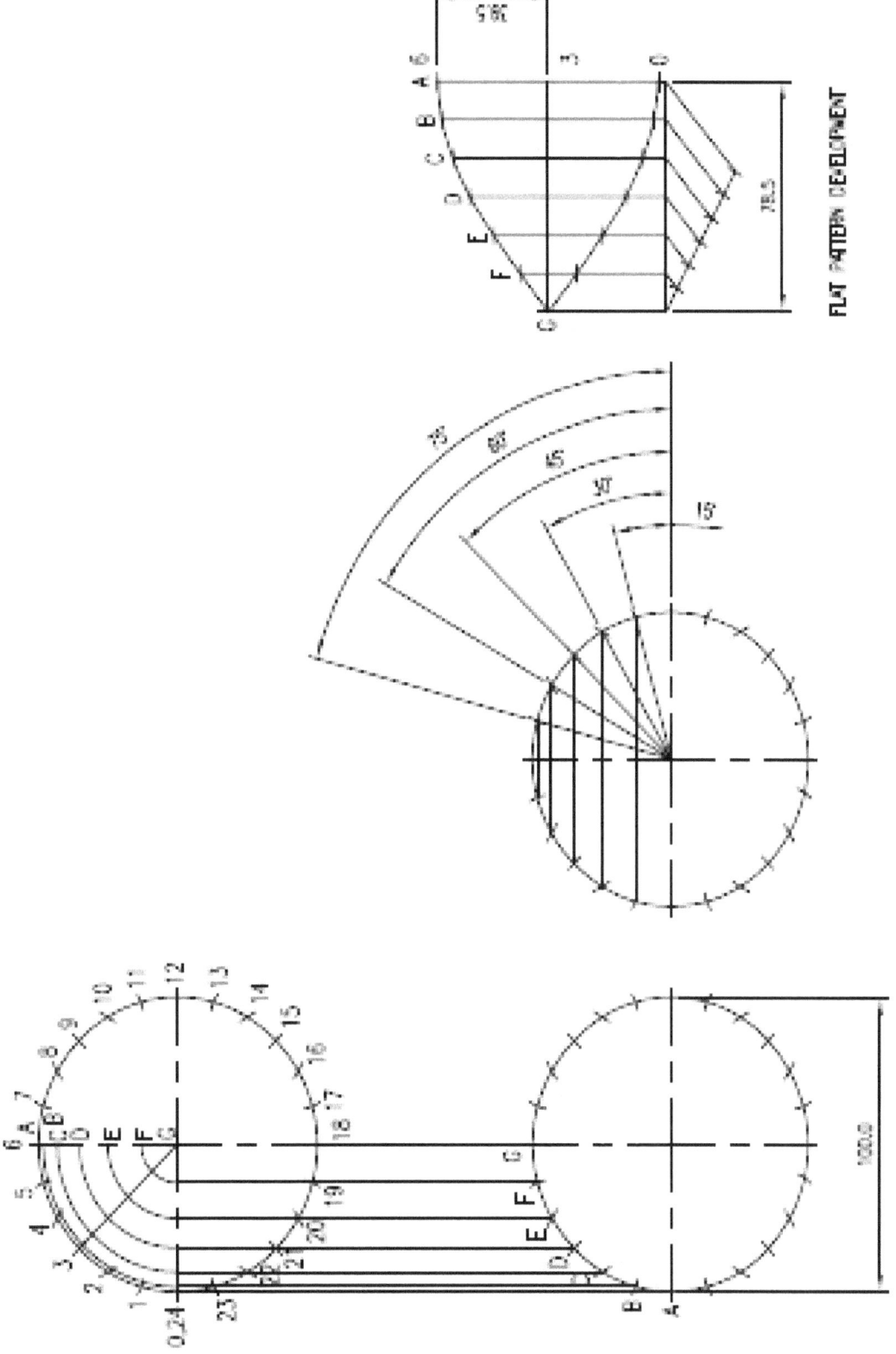

Figure 11 Development of a quarter

Development of quarter (See Figure 12)

Step1. Find quarter 24th of Ø100 = ($\frac{Mean\ Diameter\ x\ \pi}{360°}$) x 90°

$$= (\frac{Ø100\ x\ \pi}{360°}) \text{ x } 90°$$

$$= 78.54 \text{ round off } 78.5$$

Step 2. Sphere segment at 3.

At A = ($\frac{Ø100\ x\ \pi}{360°}$) x ($\frac{360°}{24}$ x 2) = 26.18 round off 26

At B = $\frac{[[[\ Cos\ 15°\ x\ (\frac{Ø100}{2})]\ x\ 2]\ x\ \pi\ ]}{360°}$ x ($\frac{360°}{24}$ x 2) = 25.288 round off 25.5

At C = $\frac{[[[\ Cos\ 30°\ x\ (\frac{Ø100}{2})]\ x\ 2]\ x\ \pi\ ]}{360°}$ x ($\frac{360°}{24}$ x 2) = 22.673 round off 22.5

At D = $\frac{[[[\ Cos45°\ x\ (\frac{Ø100}{2})]\ x\ 2]\ x\ \pi\ ]}{360°}$ x ($\frac{360°}{24}$ x 2) = 18.512 round off 18.5

At E = $\frac{[[[\ Cos60°\ x\ (\frac{Ø100}{2})]\ x\ 2]\ x\ \pi\ ]}{360°}$ x ($\frac{360°}{24}$ x 2) = 13.09 round off 13

At F = $\frac{[[[\ Cos75°\ x\ (\frac{Ø100}{2})]\ x\ 2]\ x\ \pi\ ]}{360°}$ x ($\frac{360°}{24}$ x 2) = 16.776 round off 6

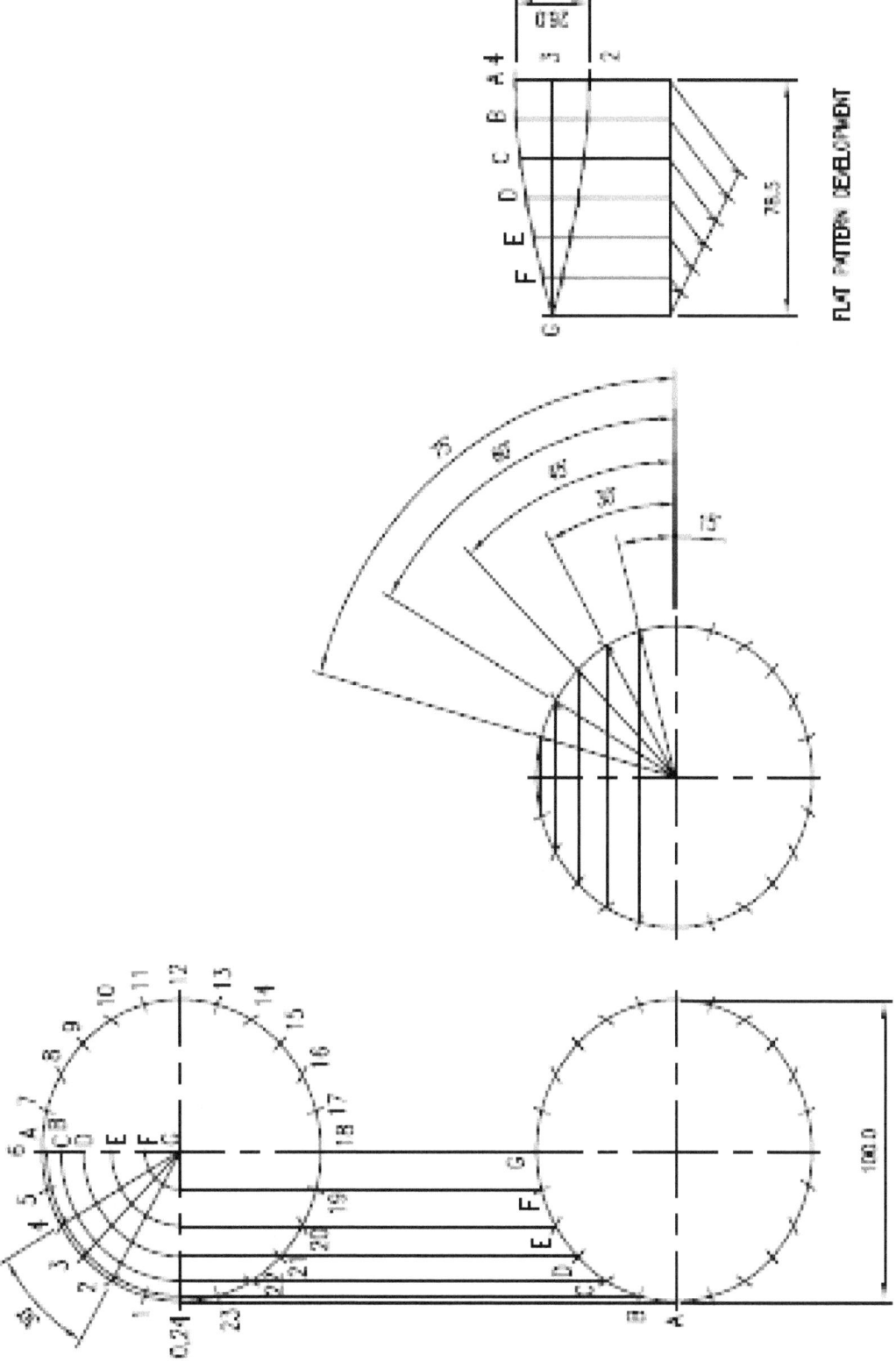

Figure 12 Development of a segment (sphere)

Development of Quarter (See Figure 13)

Step 1. Find quarter 24th of Ø100 = $\left(\dfrac{Mean\ Diameter\ x\ \pi}{360°} \right)$ x 90°

$$= \left(\dfrac{Ø100\ x\ \pi}{360°} \right) \text{ x } 90°$$

$$= 78.54 \text{ round off } 78.5$$

Step 2. Find truncated length = $\sqrt{R50^2 - R20^2}$ = 45.826

$$= \text{TAN} = \dfrac{45.826}{R20} = 66.422°$$

$$= \left(\dfrac{Ø100\ x\ \pi}{360°} \right) \text{ x } 66.422° = 57.964 \text{ round off } 58$$

Step 3. Quarter diameter half-length at 3.

At A = $\left(\dfrac{\frac{Ø100\ x\ \pi}{4}}{2} \right)$ = 39.27 round off 39.5

At B = $\dfrac{[[[Cos\ 15° \text{ x } \left(\frac{Ø100}{2} \right)] \text{ x } 2] \text{ x } \left(\frac{\pi}{4} \right)]}{2}$ = 37.94 round off 38

At C = $\dfrac{[[[Cos\ 30° \text{ x } \left(\frac{Ø100}{2} \right)] \text{ x } 2] \text{ x } \left(\frac{\pi}{4} \right)]}{2}$ = 34.008 round off 34

At D = $\dfrac{[[[Cos\ 45° \text{ x } \left(\frac{Ø100}{2} \right)] \text{ x } 2] \text{ x } \left(\frac{\pi}{4} \right)]}{2}$ = 27.788 round off 28

At E = $\dfrac{[[[Cos\ 60° \text{ x } \left(\frac{Ø100}{2} \right)] \text{ x } 2] \text{ x } \left(\frac{\pi}{4} \right)]}{2}$ = 19.635 round off 19.5

At X = $\dfrac{\left(\frac{Ø40\ x\ \pi}{4} \right)}{2}$ = 15.708 round off 15.5

At F = $\dfrac{[[[Cos\ 75° \text{ x } \left(\frac{Ø100}{2} \right)] \text{ x } 2] \text{ x } \left(\frac{\pi}{4} \right)]}{2}$ = 10.163 round off 10

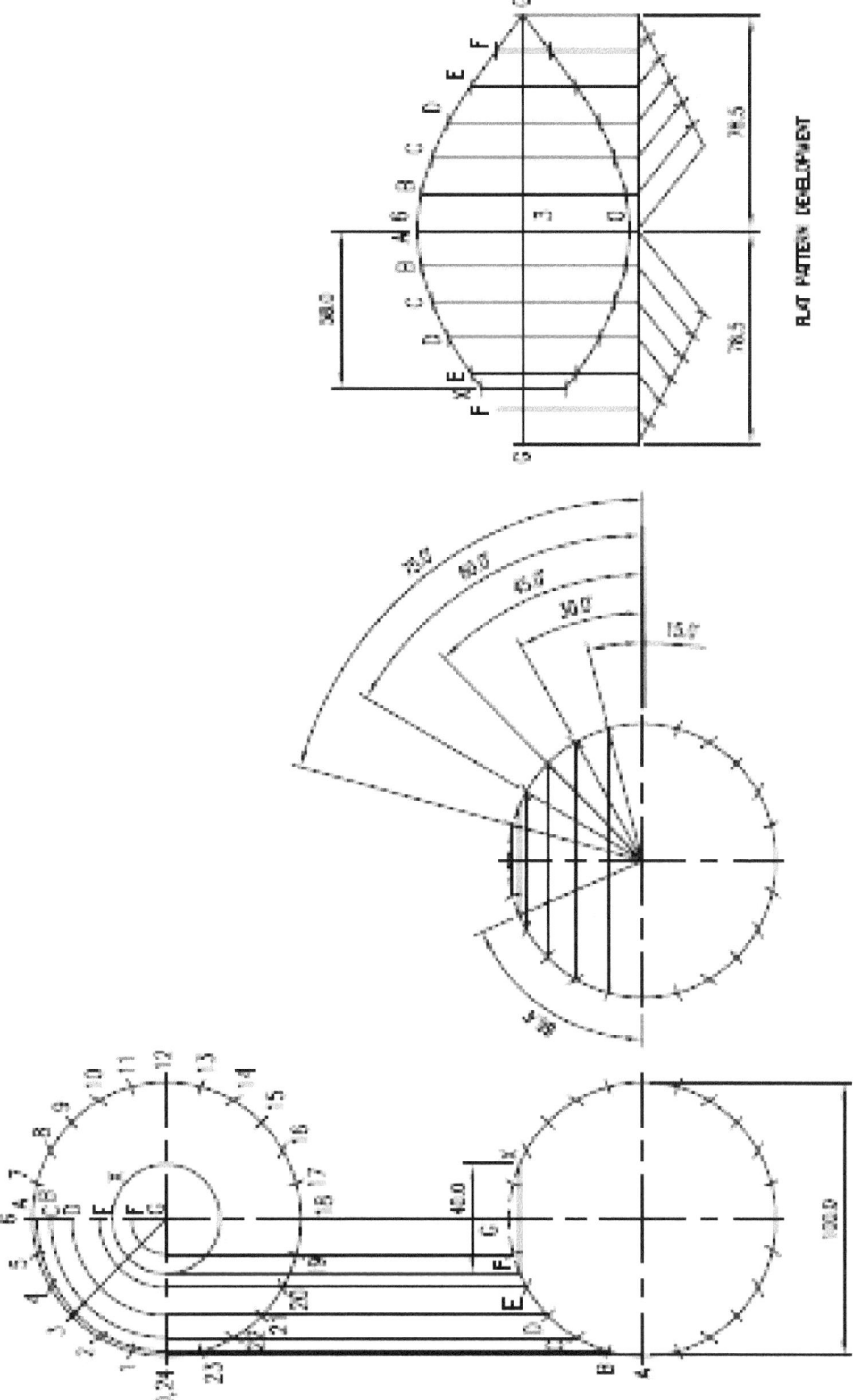

Figure 13 Development of quarter (truncated sphere)

Development of quarter (See Figure 14)

Step1. Find quarter 24th of $\varnothing 100 = (\dfrac{Mean\ Diameter \times \pi}{360°}) \times 90°$

$$= (\dfrac{\varnothing 100 \times \pi}{360°}) \times 90°$$

$$= 78.54 \text{ round off } 78.5$$

Step 2. Quarter diameter half-length at 3.

At A = $(\dfrac{\frac{\varnothing 100 \times \pi}{4}}{2})$ = 39.27 round off 39.5

At B = $\dfrac{[[[\ Cos\ 15° \times (\frac{\varnothing 100}{2})] \times 2\ ] \times (\frac{\pi}{4})]}{2}$ = 37.94 round off 38

At C = $\dfrac{[[[\ Cos\ 30° \times (\frac{\varnothing 100}{2})] \times 2\ ] \times (\frac{\pi}{4})]}{2}$ = 34.008 round off 34

At D = $\dfrac{[[[\ Cos\ 45° \times (\frac{\varnothing 100}{2})] \times 2\ ] \times (\frac{\pi}{4})]}{2}$ = 27.788 round off 28

At E = $\dfrac{[[[\ Cos\ 60° \times (\frac{\varnothing 100}{2})] \times 2\ ] \times (\frac{\pi}{4})]}{2}$ = 19.635 round off 19.5

At F = $\dfrac{[[[\ Cos\ 75° \times (\frac{\varnothing 100}{2})] \times 2\ ] \times (\frac{\pi}{4})]}{2}$ = 10.163 round off 10

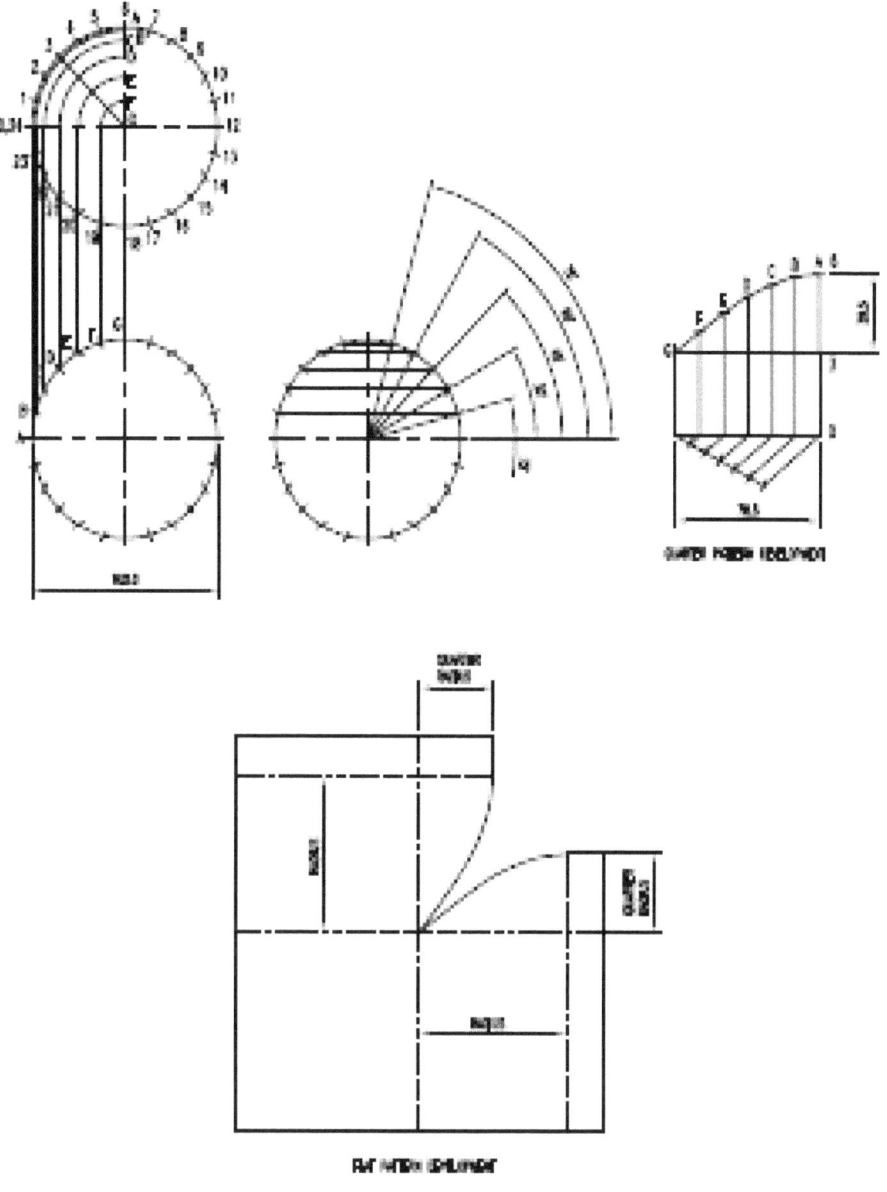

Figure 14 Sink bowl, urinal corner

REFERENCES:

1. Technical Drawing, Boundy and Hass, 1974 by McGraw-Hill Book Company Australia P/L.

www.ingramcontent.com/pod-product-compliance
Lightning Source LLC
Chambersburg PA
CBHW080644190526
45169CB00009B/3496